AFTER **OIL**

After Oil

Petrocultures Research Group

Published by the Petrocultures Research Group
www.petrocultures.com

Petrocultures
Department of English and Film Studies
University of Alberta
3-5 Humanities Centre
Edmonton, AB T6G 2E5
Canada

Contents

After Oil 2015 / Participants 7

Introduction 9

Triggering Transition 13

Energy Impasse and Political Actors 29

The Arts, Humanities, and Energy 41

Energy Futures 55

Conclusion 67

Notes 75

After Oil 2015 / Participants

Lynn Badia, University of Alberta
Darin Barney, McGill University
Ruth Beer, Emily Carr University of Art and Design
Brent Bellamy, Memorial University
Dominic Boyer, Rice University
Adam Carlson, University of Alberta
Ann Chen, University of Alberta
Ian Clarke, Ontario College of Art and Design University
Cecily Devereux, University of Alberta
Jeff Diamanti, University of Alberta
Rachel Havrelock, University of Illinois-Chicago
Olivia Heaney, McGill University
Cymene Howe, Rice University
Bob Johnson, National University
David J Kahane, University of Alberta
Jordan Kinder, University of Alberta
Richard Kover, University of Alberta
Ernst Logar, Independent Artist
Graeme Macdonald, University of Warwick
Negar Mottahedeh, Duke University
Michael O'Driscoll, University of Alberta
Fiona Polack, Memorial University
Sina Rahmani, University of Alberta

Jerilyn Sambrooke, University of California-Berkeley
Jackie Seidel, University of Calgary
Mark Simpson, University of Alberta
Lucie Stepanik, University of Alberta
Janet Stewart, Durham University
Imre Szeman, University of Alberta
Kevin Taft, Independent Scholar
Michael Truscello, Mount Royal University
Aaron Veldstra, University of Alberta
Carolyn Veldstra, University of Alberta
Caleb Wellum, University of Toronto
Sheena Wilson, University of Alberta
Saulesh Yessenova, University of Calgary

Introduction

"After Oil: Explorations and Experiments in the Future of Energy, Culture and Society" is a collaborative, inter-disciplinary research partnership designed to explore, critically and creatively, the social, cultural, and political changes necessary to facilitate a full-scale transition from fossil fuels to new forms of energy. A foundational prem-ise underpins the work carried out by the "After Oil" research team: energy plays a critical role in determining the shape, form, and character of our daily existence. The dominant form of energy in any given era—in our case, fossil fuels—shapes the attributes and capabilities of soci-eties in a fundamental way. Accordingly, a genuine global transition away from fossil fuels will require not only a reworking of our energy infrastructures, but a transforma-tion of the petroculture itself.

What do we mean by "petroculture"? We use this term to emphasize the ways in which post-industrial society today is an oil society through and through. It is shaped by oil in physical and material ways, from the automobiles and highways we use to the plastics that permeate our food supply and built environments. Even more signifi-cantly, fossil fuels have also shaped our values, practices, habits, beliefs, and feelings. These latter can be difficult to parse. It might be easy to point to a highway interchange and understand its relationship to our oil culture, but it is much harder to name and isolate the ideals of autonomy and mobility, for instance, that are just as strongly linked to the historical conditions of a fossil fuel society. In a very real way, these values are fueled by fossil fuels, as are so many of the other values and aspirations that we have come to associate with the freedoms and capacities of modern life. It is in this sense that we are a petroculture; and it for this reason, too, that transitioning from fossil fuels to other sources of energy will require more than new energy technologies. We will need to transform and transition our cultural and social values at the same time.

In August 2015, thirty-five artists and researchers came together in Edmonton for the inaugural After Oil School (AOS). They were invited to think collectively about the challenges living in a petroculture poses for energy transition. Over four days, they were asked to discuss, debate, and to provide answers to four key questions:

1. Considering historical precedence, what cultural strategies are available to trigger and expedite a large-scale transition of energy regimes?

2. How does the problem of energy force us to rethink our traditional notions and categories of political agency?

3. How is the use of energy entwined with representations and narratives about modernity and the environment? Correspondingly, how do artistic productions reflect, critique, and inform our understanding and use of energy?

4. What range of scenarios is currently on the table for imagining our future with energy?

This short book includes the answers to these questions, organized in sections that correspond to the order of the questions above:

1. Triggering Transition

2. Energy Impasse and Political Actors

3. The Arts, Humanities, and Energy

4. Energy Futures

The first chapter explores how we might begin the process of energy transition through social transition, concluding with a set of principles for an effective, intentional energy transition. The second elaborates the most common narratives that we have about our fossil fuel society and the forms of political action that are set out in each of these narratives. These varied understandings of how we define the problem of fossil fuels and a transition from them gives us an insight into the multiple levels at which political action will need to occur for a genuine transition to take place. The third chapter describes the unique critical capacities of the arts and humanities in making sense of our petrocultures. Finally, we reflect on energy futures and consider how looking ahead might help to lead us to a new kind of society—one for which it would no longer make sense to use the term "petroculture." These chapters can be read on their own or as contributions to a larger argument about all of the issues and problems we will need to consider as we try to move to a time and place after oil.

One of the many things that make this short document distinctive is that it is a collective document, the product of intensive work by thinkers committed to addressing the difficult questions we will need to pose—and answer—if we are to ever get to a world after oil. It is this kind of collective work that will be needed over the coming years and decades to transition from fossil fuels to renewables, and from a petroculture to the new global culture that we can see just over the horizon.

—Imre Szeman, Lynn Badia, Jeff Diamanti,
Michael O'Driscoll, and Mark Simpson

Organizers of the 2015 After Oil School
(August 19–22, 2015)

www.afteroil.ca / www.petrocultures.com

Triggering Transition

Energy transition is social and historical: the history of energy expresses a complex set of social commitments that develop over time. Triggering transition in the present means engaging in that history and those relations.

Energy transition is not simply—it is not even mostly—a question of technology and the economic policy of supply, although it is also, of course, that. The energy question is, at its core, a human question, a social question that concerns accounting for the quality of human experience under the fossil economy, reckoning with the increasing precarity of life under fossil fuels, and seizing the opportunity to redress the failures and the blocked desires sedimented in the old economy. The energy question centres on the values that frame our lives and the possibilities for a quality of life that might be made available to us by decoupling ourselves from petroleum, natural gas, and coal. Yet the epistemological and political recognition that energy transition implies (and might well be implied by) social transition does not immediately *trigger* transition. If it did, the first Intergovernmental Panel on Climate Change (IPCC) would have sealed the deal; we'd find ourselves today in a world firmly *after* oil. We don't. The trigger—that historical, intentional set of forces that actualizes energy transitions—is not reducible to the hard facts of transition itself.

Fossil fuels have made possible the greatest era of social, technological, and economic growth this earth has ever seen. Oil, likewise—and importantly due to its growth-giving capacity—has generated its own logical, physical, and social impasse. *After Oil* begins by taking these two sides of oil as central to the concept and challenge of energy transition. The "after" in *After Oil* thus refers both to the consequence of oil, since we live in a world contoured by a fossil-fuelled modernization process, and to the opportunity of transitioning to a world where fossil fuels no longer dominate our energy systems.

A transition that meets basic human needs and reflects collective desires requires a social framework. There is no shortage of positions that indict, expose, or politicize oil and fossil fuels. And for good reason. Rapid environmental degradation and the now incontrovertible evidence that we are in the midst of an epochal transition in climate patterns occasion a good deal of alarm, confusion, and anger. Fossil fuels are now thoroughly politicized. Industry and progressives, privileged consumers and the disfranchised, battle it out in the streets and in the media with radically unequal resources. But the humanistic project of reframing energy as a social or human question has not advanced very far. Currently, new energy inputs such as wind power, solar power, biofuels, and so on are posited as the endgame of the transition, but fundamental commitments to values, to satisfying social relations, and to our collective imaginaries are, at best, left to the margins of the discussion, if not erased from the conversation. Establishing a new social framework is not merely a question of policy or financial investment. To imagine a society after oil means first understanding what oil is to us—how it shapes current desire, identity, and practice, comfort and pain, consumption and penury.

There have been previous energy transitions. There have been social transitions. However, there has never before been a transition demanded of us, and on this scale, that requires such forethought. The only historical transition that gives us insight into what is on the horizon (i.e., the scale of infrastructural and social shift) is the transition into the energy and economic system we're on the brink of exiting. This is the epistemological and practical problem of the impasse of fossil fuels—that is, what blocks us from transitioning to other forms of energy—and of the economy locked into its rhythms.

What is Impasse?

We take it as self-evident that we are at an impasse like no other in history. Without signposts, we now must transition to different ways of being in the world, both with each other and in relationship to the environment. In this context, the direction forward is not preordained or written into the problem. While many of us remain optimistic that we can sustain our attachment to oil and the good life that it has come to define in the global West, it is increasingly clear that a continuance of the fossil economy is a form of "cruel optimism" that not only carries forward old risks but also introduces radically new risks into our lives.[1] We now know, deep down and viscerally, that oil is problematic. Reckoning with that fact requires lucid analysis and imagination. Thus part of the work of transition is to make visible our social, material, and affective attachments to oil: to its role in the social and cultural formation of our everyday lives, the infrastructures and institutions of our social interconnectedness, and global networks of relations.

The transition to a society after oil is stalked by the experience of impasse. Oil is so deeply and extensively embedded in our social, economic, and political structures and practices that imagining or enacting an alternative feels impossible, blocked at every turn by conditions and forces beyond our understanding or control. Impasse, understood in this way, invites paralysis and reinforces the status quo.

But what if we were to think impasse otherwise? Rather than understanding impasse as foreclosure of possibility, we posit that impasse is a situation of radical indeterminacy where existing assumptions and material relations can no longer hold or sustain us and in which we might activate the potential obscured by business-as-usual. In this case, an impasse is not a blockage; it is a condition of possibility for action within a situation that is suddenly open because it is uncertain. Impasse is, in other words, a moment for aspiration and courage. This moment is the transition to a society after oil.

To reiterate, impasse can be an optimistic space, a liminal space, a space of hope in which we can attempt on many different levels and social registers to begin to articulate the outcomes of less energy-intensive lifestyles. While the new ways of being in relationship to energy, the environment, and one another will be built on the legacies of oil, there is the opportunity for breaking with the limitations in that legacy. The current moment thus provides us the opportunity to think through what the age of oil brought us, what we want to salvage and maintain, and where we want to work to construct more equitable and just social relations in the age to come: after oil.

What is Oil?

Oil composes space and shapes culture. It modulates our

lives, including the clothing we wear, the objects we use, the buildings we occupy, the spaces we move through, the daily routines that structure everyday existence, our habits and perceptions, our commitments and beliefs. Oil (as a metonym of the larger fossil economy) is, in other words, not just a substance one pumps into the car. And nor can it be reduced to the abstract figures that rise and fall in the financial pages of the daily paper. Oil names a way of organizing society, of bringing people together, and of keeping them apart.

Put another way, oil is not simply a source of energy: mere fuel, brute input. It is inextricably social.

To describe oil in this way is to view the problem of energy transition from an unfamiliar perspective: not simply as the site of a new technical difficulty that must be resolved but as the object of a social challenge. For to transition from oil to some other energy source will entail—whether we like it or not, whether we participate in the process or opt out—the unmaking and remaking of our social worlds. Undeniably, this prospect is daunting, even overwhelming. But might its challenge also offer surprising promise and possibility?

The reason, however, that oil modulates everything is not some natural or magical property of the energy source itself. Rather, oil expresses a social system bound up historically with the rise of modern industry and industrial capital, including the creation of an industrial working class (now barely visible from within centres of advanced economies); the birth of middle-class opportunity and material privilege in the West; and the mirrored acceleration of precarity and mass unemployment across the globe. Energizing the labour process at the site of production increased the productive capacity of workers, but it also gave business owners a solution to the rising cost of labour. Today, we call these phenomena automation, offshoring, and capital deepening, yet as economic strategies all three depend on more and more non-human energy in the form of transportation and more efficient machinery. These phenomena make visible the relation between reducing labour costs and increasing dependency on energy outputs in a formulation known as "energy deepening."[2] Read from the standpoint of oil's industrial beginnings, rising unemployment and economic disparity are logically consistent with a specifically fossil-fuelled form of capital.

In the long view, pairing human labour first with coal power and then with oil's uniquely dense, powerful, and volatile properties has overcome material and seasonal constraints while causing new and much larger environmental constraints. Economic crisis begets environmental crisis, since consolidating economic power in the hands of the few has been achieved through energy deepening, just as environmental degradation implies a rising volatility in the economic sphere, since energy deepening implies labour shedding. The history of oil is the history of the present. An intentional transition away from fossil fuels will need start by attending to the deep links that have been forged between profits and global warming, GDP and CO_2.

What is Intentional Transition?

The self-evidence of oil's social embeddedness and the need for energy transition requires an assertion of agency, a conscious seizing of the opportunity presented by today's impasse.

If oil so saturates our cultural and social imaginary, then what is one to do? What options are available to us in the midst of this tectonic transition that is moving underneath our feet and circulating in the air we breathe? Given that we are already deep in the midst of transition (if not an intentional, focused one), where should we locate ourselves? The default position is a disabling one. It is to assume that this transition is a purely technological problem that will be resolved through technocratic solutions. Such a position assumes that responsibility can be entrusted and handed off to someone else. Reinforcing this default resignation is the embedded assumption that the market will resolve the crisis. This, too, presumes that the only intentionality needed is that of market forces, and that we, as individuals and communities, need not participate in moulding, shaping, hoping, or imagining, except along narrowly defined consumerist lines. To accept this default position is to abdicate agency. It is to abandon to someone else the creative act of making the world and the values that it will hold.

An intentional transition reframes the energy question as a humanistic one requiring our vote in the matter— our intentionality, agency, and the assertion of values and desires that we hold. As such, it begins by taking account of where we sit historically, where we find ourselves in terms of our infrastructural dependencies and our affective and erotic attachments to the fossil economy. An intentional transition begins by reckoning candidly with the problem of the path dependencies that are required for survival in a post-oil economy and with an acknowledgement of the attachment to desires realized under the fossil economy. But it then moves beyond oil to a reckoning with the failures—the blocked desires—the pain and penury, the inequality and injustice, which the fossil economy could not resolve under its terms of management.

Triggering

What is a trigger? The dictionary answers in technological terms: the lever one pulls to release a catch, fire a pistol, or spring a trap. This answer and the images it conjures are vividly straightforward. They emphasize mechanical action: the comforting simplicity of cause-and-effect. When the issue in question is the wholesale transition in the mode of energy that powers our world, from oil to some other form, a simple, mechanical answer can seem incredibly seductive. But its suitability, its explanatory power, is limited. This answer (a lever that initiates an action, a cause that results in the blink of an eye in an effect) is itself a trap. We need to understand the trigger, and triggering, otherwise.

One way to name and so grasp the trigger for energy transition in the present is the global warming caused by human-induced climate change. We know this version of the trigger intimately even as we disavow it relentlessly; this trigger triggers our most dread-laden nightmares of incomprehensible future catastrophe. Global warming as trigger also clearly complicates the mechanical view supplied by the dictionary, since the sense in which humans have pulled this trigger completely undoes any ordinary sense of what pulling means. Global warming as a trigger for energy transition constitutes something like a forced choice: shift to a sustainable form of energy, or burn out the planet.

The environmental trigger for energy transition is certainly compelling. But it bears on the problem of transition along only one axis: with regard to fuel source, yet not necessarily with regard to social form. This result impoverishes our understanding by luring us into the mistake of imagining energy as prior to and distinct from the social.

A recognition and engagement with the deep inextricability of energy and society, by contrast, will require—but perhaps can likewise enable—a perspective on triggering that is adequate to this inextricability.

The coal-powered industrialization of English manufacturing in the nineteenth century sparked the largest energy transition in human history. Received accounts of the rise of modern industry (the process familiarly called "The Industrial Revolution") typically associate the adoption of coal powered steam engines with a straightforward increase in productive capacity and efficiency: in other words, with a clear narrative of technological progress. In this account, technological determinism is both the trigger and the transition: some innate urgency to increase efficiency and output triggers the transition to new energy inputs autonomous from the social history that works in and consumes the products of modern industry. Social history, in this linear version of progress, is an expression of technologically driven economic growth. This same historiography is echoed today in promises that the market will naturally select the most environmentally and economically efficient solutions to climate change. *Homo economicus. History technologicus.*

Recent work by the social historian Andreas Malm makes a compelling case for a different way of understanding the emergence of the fossil economy.[3] By Malm's account, the shift to coal in industrial manufacture occurs decisively in Britain's cotton industry in the 1820s and 1830s despite the fact that, at that moment, water remains a considerably more potent (and cheaper) source of power to drive industrial machines. Puzzled, Malm asks why factory owners make the switch to coal if water was both cheaper and more efficient. Viewed strictly technologically, it makes no sense. Viewed socially and economically, however, it does: switching from waterpower to coal power meant that factory owners could move production into dense urban settings where workers were numerous and cheap. Coal simultaneously intensified and regularized the ten-hour workday, and liberated factory owners from the spatial limits of waterpower. In cities, more labour could be exploited at higher levels of intensity. In effect, fossil fuels triggered the industrialization of both machine power and labour power, enabling cotton capitalists to solve the falling rate of profit and to circumvent—or indeed sabotage—the nascent power of organized labour by turning to the unemployed and so driving their production costs down.

As a way to comprehend the trigger for energy transition along two axes—social relations as well as fuel source—Malm's case is both vivid (since it dramatizes the inextricability of energy from society) and discomfiting (since it hardly offers a model to replicate). Will a global unemployment crisis trigger a renewables revolution? Will market driven technological determinism pick an environmentally sustainable mode of production? Actually, we might answer both in the affirmative and still wonder whether the previous trigger—the need to more efficiently and consistently exploit increasingly hostile bodies of labourers—is one we are willing to endorse today. In any event, Malm's lesson remains instructive, precisely by indicating the priority of social and economic questions and relations for any transitional trigger out of the fossil-fuelled energy world we continue to inhabit. To grasp the trigger today, in other words, means first grasping the social relations we have and, even more urgently, working to propose and then to materialize the ones we might want.

Acting into the Impasse

To act during an impasse takes courage. This is especially true of an impasse experienced as an occasion, a site of contingency, and a moment of possibility, in which the outcomes of acting cannot be guaranteed in advance. After all, it is this very indeterminacy that turns an impasse into a political situation. This is also why politics provokes such reticence. Politics is predicated on a disturbance in the status quo. Our typical response to such disturbances is to make action conditional upon an assurance about how things will be when the situation is resolved. It is this response, and not the impasse itself, that drains the situation of its potential.

The transition to a society after oil means more than just finding a replacement for fossil fuels that will allow all the social practices and relations bound up in our current energy regime to remain as they are. Aspiring to a society after oil means that these practices and relations will change. Acting into the impasse of oil means getting down to the work of remaking social practices anew under conditions in which we cannot be certain of how things will end. How will we pay for our schools if the oil companies no longer extract the resources below the ground? We don't know for sure. But this is where we must begin, right here in our present practices and institutions, some of which will be transformed, some of which we might have to leave behind altogether. But we will never act so long as we are discouraged, so long as we insist on the end before the beginning. If we already knew the end, and we already knew how to install it with certainty, then we would not be at an impasse, and there would be no need to engage in political action.

Those who profit disproportionately from the society of oil are happy and quick to discourage us. But being discouraged is a luxury we can no longer afford. Encouragement at the impasse is what the humanities can provide in the transition to a society after oil, not because these disciplines foretell the future, but because they open us to a thoughtful and responsible composure towards its uncertainties and possibilities. They teach us not to fear difference when we can no longer retreat into the same.

Energy Deepening

Energy deepening names the tendency through which capitalist modernization mobilizes natural forms of physical power to optimize, manage, and discard human labour. Without rising levels of productivity from employees, business owners cannot retrieve profit in a competitive marketplace. Without quarterly expansions of national economies, state and municipal budgets flat line. One solution to this fact of economic life has been to bring more and more workers into the workspace in order to stimulate cooperative output (manufacturing). A second has been to invite cheaper labour into the marketplace, or to search it out elsewhere (offshoring). Another has been to pair workers with more and more energy-hungry machines fuelled on coal and then electricity (industrialization). A fourth strategy, more familiar to the recent experience of postindustrial societies, has been to replace workers with technologies able to do the same job (capital deepening). All four strategies, however, depend on a steady rise in energy inputs further and further removed from the spaces of labour.

The global marketplace is another name for the spatial result of energy deepening, since decades of cheap oil prices made possible the logistical and communications networks that globalized the economy and its geographical distinctions. This, in short, is how oil generates the setting of the global marketplace, in addition to its social, material, and cultural content. So long as the time and space of oil is taken *as the world*, the transition to a world after oil will remain categorically impossible. Once oil's role as a modulator of economic and thus social relations is brought to the centre of the project of transition, the stakes, content, and form of what is in transition alter dramatically. This is the drama *After Oil* takes as empowering.

The sequence initiated by the industrial revolution depended on the economic necessity of energy deepening. The transition out of that sequence will—of social and ecological necessity—make energy deepening unnecessary.

Principles of Intentional Transition

First, agency and mobilization

An intentional transition is premised on agency, on the conscious participation and mobilization of peoples and communities. In this respect, conscious participation cannot be reduced to the meagre practice of constituencies being brought into a discussion after the terms of the debate have been set. It means people being brought together to establish the framework for debate from the start, so that its terms and its conduct conform to their hopes, their needs, and their values as individuals, families, and communities.

Second, collective stewardship

An intentional transition is premised on collective stewardship, on the avowed right of people and their communities to own, manage, and develop the energy resources that conform to their desires and needs, and that support their ideals for reproducing and producing the health of their communities and the values they hold. In this sense, public control is distinct from the prevailing tendency toward private control and increasing private management of this epochal transition.

Third, equality

An intentional transition is premised on equality, on the right of all peoples and communities to adequate energy resources for survival. It is to acknowledge that life under the fossil economy did not fulfill for many people or communities this basic human right, and that the fossil economy produced wild inequalities that left much of the world behind while conferring the privileges of energy along unfair, and wholly undesirable, racial, national, gender, and class lines.

Fourth, ethics of use

An intentional transition is premised on a clearer understanding of the ethical dimensions of energy use and the hierarchy of human priorities. Intentional transition means collectively sorting out the moral differences between the use of energy for the more elementary needs we all have for food, water, and the basic essentials of life, and the surplus material and immaterial desires that energy quite literally feeds and fuels (more on transition desire below).

Fifth, sustainability

An intentional transition is premised on sustainability. It distinguishes quite clearly between accepting the risk of an increasingly obsolescent fossil economy and embracing the opportunities of an after-oil economy in which energy is thoroughly socialized and generated within a framework of sustainability. To that end, it assigns renewable alternatives a central place in the transition away from those dependencies that have produced climate change and the current culture of risk.

Sixth, redefinition of growth

An intentional transition is premised on growth and development. But, importantly, it does not take these terms as self-evident. Instead it redefines these much-abused terms as something distinct from business-as-usual. In the after-oil economy, growth and development are tied to the social values articulated above and joined to a new ethics of resilience and sustainability. Growth and development are taken out of the hands of the economists and given back to the people.

Transitioning Desire

Some of the challenges involved in intentional transition can be grasped by considering just one of its many dimensions: shifts in how desire is coordinated by and in relation to the use of fossil fuels.

In the Western world, we live in an era of unmatched material plenty in which desires are indulged and encouraged, no matter how apparently trivial. A consumerist ethos pervades our culture and for many it appears that we inhabit (in the words of former American President Herbert Hoover) the world of the "constantly moving happiness machine." The incredible cornucopia of the twentieth and early twenty-first centuries would have been unthinkable without a cheap, portable, seemingly infinite source of energy in the form of petro-carbons or oil.

As we have seen, our dependence on oil has had unforeseen but profoundly dire consequences to the ecological health of our planet that, if unaddressed, could prove catastrophic to both our natural and social worlds. Attempts to address this crisis have largely concentrated on advocating transition to more "renewable" forms of energy, yet as critics such as Vaclav Smil point out, it is unlikely that, now or in the foreseeable future, these forms of renewable energy will be able even to supplement our current energy demands, let alone those of the future, which are likely to be far greater.[4] Our present circumstances amount in part to a crisis of desire whose resolution may depend less on finding new, less ecologically destructive forms of energy, than on restraining or curbing what looks to be a limitless desire provoked and fuelled by consumerism. Such a formulation sits uneasily with the modern temperament, and, in the face of promises of unrestrained plenty, the suggestion of restraint smacks of puritanical sanctimony and invites such questions as "Who are you to tell me to forego my desires?" Nevertheless, tackling the question of desire need not require the suppression or even renunciation of desire but rather, as Yannis Stravrakakis has argued, its redirection.[5]

If life in consumer society promises a dream of endless ease and joyful satiation, its critics have often pointed precisely to the profound gap between this dream and actual lived experience, noting that the actual pleasures and happiness experienced fall far short of those promised. To such critics, the consumer citizen appears very much akin to a dog chasing its own tail, pursuing an elusive goal that it can never achieve, no matter how fast it runs. Given the frequently noted intimate connection between petroleum as primary energy source and the deterritorialization, intensification, and acceleration of production, it is to be wondered whether the transition from fossil fuels might itself offer new opportunities to satiate human desires for things as a more intimate connection to local social and natural communities, fulfilling work and free time.

Energy Impasse and Political Actors

Oil is not only something you put in your car. It is the foundation of our political identity and institutions, and it profoundly shapes our society and environment. But how we tell the story of oil, both of its past and its possible futures, shapes how we see (and perhaps also whether we see) the problem at its core. An impasse is a situation in which progress is not possible due to entrenched disagreements or deadlocked opinions. Structural features also contribute to the political blockages barring routes to a post-oil world. Carbon reliance, a capitalist economic system, and climate change are just a few of the factors combining to generate the current political impasse around energy. The stories we tell about our energy use each frame this impasse differently, and in so doing, also identify different routes out of it.

We've identified six different narratives we tell about oil's past and how we might transition out of an oil-based world:

1. Transition from Below

2. Transition Without Loss

3. Transition Through Localization

4.   Transition after Capitalism

5.   Transition Through State Reform

6.   Transition Through Catastrophe

Transition from Below

What's the story?

To achieve the necessary transition (in energy, but also away from unjust and alienated social relations) we need to build alternatives together that use resources more sustainably, (sometimes) involve new forms of energy, and build alternative understandings of wellbeing (not premised on consumerism). This may involve confrontation with dominant state and corporate forces, but these are not our political focus. By building alternatives for ourselves we're building new forms of community and overcoming disempowering forms of alienation in favour of solidarity and human relationships.

Who tells this story?

Permaculturalists, some indignadas movements, activists and citizens involved in direct action resilience, transition towns, and more. Though interestingly, many of the people building a different energy future with their own hands—whose work is at the heart of this story—may not articulate this story; they simply live it.

What's the impasse?

The massive power of energy corporations and the complicity of dominant political and economic institutions in our current energy system. Along with this comes a sense of disempowerment among individuals and communities that is created through lives that are pressured economically and are marginalized. Too many of us, out of our marginalization and sense of disempowerment, have little sense of political efficacy and may focus on consumption rather than on community or our contributions to others as our sources of wellbeing.

What is the pathway to action?

Empowerment is created by building the alternatives we need not only as individuals but also in community. To achieve this we need to develop an economy that enables new forms of collaboration and action, and which recognizes the need to support people through the psychological and existential challenges of transition. We need to connect experience across the levels of individual, household, and community; in other words, we need to build a collaborative society based on alternative models of social and economic organizations and a sharing of diverse skill sets, knowledges, and experiences. Some groups following these kinds of approaches have experienced evident rewards, but a challenge remains in making these more widely visible and achievable.

Lingering questions?

Can a hands-on, non-hierarchical model flourish at a larger scale in the current context (where it would have to interact with state mechanisms or corporate players)? Or does this story rely on the collapse of the existing system before it grows? Is this story inevitably heard as forlorn or naive in the face of a rapacious, highly resilient dominant system? How can this story compete with the lure of conventional models of success based on upward mobility? How does this mode of communal organizing address differences in ability, resources, and social location?

Transition Without Loss

What's the story?

This transition imagines a wholesale conversion through decarbonizing the current economy, through new technologies and/or a switch to renewable energy sources without the loss of basic structures of life; indeed, this story often imagines an improved quality of life for many, if not most of the planet's inhabitants. There's a subset of this argument that emphasizes gains—extending capitalism's "green" growth through new energy technologies.

Who tells this story?

Energy corporations, governments, and technology companies.

What's the impasse?

This story relates the impasse of energy quite simply: we don't have the right technology in place yet.

What is the pathway to action?

This story's narrative sees current leaders and decision makers buying into new energy systems and transforming the market through education, subsidies, and regulation. Technocrats—those with access to the knowledge and funding necessary to build new energy systems—are at the heart of deciding what a non-carbon infrastructure will look like.

Lingering questions?

Without loss for whom? This narrative fundamentally points to a different energy system, so the "without loss" idea is disingenuous; this bleeds into an anti-capitalist model quite quickly, since the capitalist model is the one that fosters a culture of scarcity and competition that requires loss on the part of some. This "risk-free" path to a solution risks retaining existing socio-economic inequalities—indeed, requires them—and has the potential to create zones without access to resources for those without the capital to put new systems in place. We might ask: how does this story address the current capture of the state by players in carbon energy? How does this story propose to address the rampant inequality and injustice that are central to capitalism? Finally, can we reimagine loss in order to reframe this narrative? Loss is only imagined in terms of a commodity system that imagines a consumer at its core. Can we embrace certain kinds of loss? How would we do that? How would we encourage others to do that? Could we balance losses as a form of gain in other terms?

Transition Through Localization

What's the story?

The existing allocation of resources to corporations is inefficient, exploitative, unjust, and ecologically damaging. This mode of transition envisions shifting energy management, ownership, and allocation away from corporations and towards a system of common, or shared, resource stewardship among people living in a particular place. As the impacts of climate change begin to affect more people in terms of drought, flooding, heat waves, public health epidemics, the timing may be right for a re-examination of who benefits and who pays for the effects of massive carbon release.

Who tells this story?

Indigenous communities often promote stewardship as a

component of community. This is also a narrative told by activist groups seeking to establish regional renewables companies. Communities whose water sources are contaminated by energy extraction often advocate for greater local control and oversight of the water-energy nexus in their region.

What's the impasse?

Energy has seldom been viewed as owned or managed by those who live near energy sources and its infrastructure, and so few examples of place-based or local ownership exist. In addition, corporate ownership and profits are protected by current juridical, political, and economic systems so as to make local ownership and management nearly impossible.

What is the pathway to action?

Developing a transition through localization of resource management by promoting a politics of presence and resource stewardship, or, developing the idea that those who live in a given region have a stake in the management of local resources. This transition also depends on making successful examples of local ownership and generation more nationally and internationally visible.

Lingering questions?

If energy could be turned into a commons, what would that look like? How does this narrative address profound differences in access to energy/water resources in different places? How can an approach based on localization become transnational or global? How would such an approach operate in spaces like offshore drilling platforms? How would this approach entail community driven institutions? How would this approach deal with the historical disenfranchisements of local populations? How can corporations be made accountable to the local effects of resource extraction? What does energy dispersion and use look like outside of a profit model?

Transition after Capitalism

What's the story?

Capitalism is growth-oriented and accelerationist at its heart—that is, premised on intensive and extensive gains—and therefore at odds with a transition towards reduced energy use. Only by breaking a much broader system of capitalism can we achieve transition out of carbon-based energy reliance. In other words, there is an intrinsic link between justice struggles and energy transition.

Who tells this story?

Naomi Klein; Kolya Abramsky; Midnight Notes Collective; Anarcho-Primitivists; Communitarianians and Utopian Socialists; Marxist ecologists; proponents of World-Ecological theory.

What's the impasse?

Industrial capitalism has been powered since its beginning by fossil fuels; you can't change the problem of energy without changing the system. Yet it's difficult (impossible?) to imagine a life other than that produced through capitalist means. The impasse, then, is the immense appeal of our oil-based lives and the weight of the physical and social infrastructures produced over the life of oil. And let's not forget, too, the massive power of corporations—who are inclined to preserve the status quo— over individuals.

What is the pathway to action?

This narrative imagines its heroes as "individuals of conscience" prepared to stand up to the systemic agency of capitalism. An activist approach to confronting oil capitalism seeks to mobilize citizens against the state and corporations through social media campaigns, education, divestment campaigns, solidarity building, and/or direct action, and to persuade workers to realize the value in the jobs and egalitarian opportunities of alternative energy infrastructures. Yet the very systemic power and agency of states and corporations also makes it difficult for this story to really believe in its hero.

Lingering questions?

The strength of this story lies in its critique. How might we translate that critique into meaningful systemic change? And what do we make of this approach's tolerance for violence (even if, to date, there has been minimal violence in the name of energy transition)?

Transition Through State Reform

What's the story?

This model of transition imagines large-scale state intervention that can range from a slow-paced reformist and regulatory approach to a large-scale rapid and radical reorganization of space and resources.

Who tells this story?

Politicians invested in social change; NGOs; international organizations and governance structures; authors such as Kim Stanley Robinson (as in his Science in the Capital trilogy).[1]

What's the impasse?

The state has been made subservient to the economy and in many cases the state has grown up with/on the carbon economy, so it's difficult to see how the state could be uncoupled from corporate interests/capitalist economy.

What is the pathway to action?

This story imagines politicians and political parties working to transform approaches to energy on a wide scale through existing political and juridical processes. Corporations are seen as innovators in this process, as they enact internal transitions in compliance with state reform. Alongside this, civil society acts as a "policing" force to ensure the state's role and actions in energy transformation. Supranational organizations such as the International Monetary Fund or World Trade Organization, as well as international trade agreements, work to engage transformation worldwide.

Lingering questions?

Can real transformation be achieved through an approach that reinforces a capitalist model and commodity view of energy? This story entrenches the state as a vested interest in the carbon economy, which begs the question of whether the state could survive a transition to renewables. What are the outcomes of centralizing energy resources in global geopolitics? Does this approach invite state-sanctioned violence, surveillance, displacement, and disenfranchisement? How can energy workers be convinced of their vested interests in an after oil scenario?

Transition Through Catastrophe

What's the story?

This is the story that tells us that we don't really have the ability to comprehend what awaits us after the end of carbon democracy. The reason for this is that our socio-political institutions and even categories of social analysis (e.g. "growth" as measure of economic health; "base load" as an expectation of grid logistics) are so deeply embedded in the logics of fossil fuels that we cannot imagine what a post-carbon society would look like. Ironically, in some apocalyptic narratives, a world "after oil" is envisioned as inherently "catastrophic," thereby providing a convenient argument for the status quo. This is the root cause of our present condition of impasse. The implication is that some kind of rupture, possibly catastrophic, would need to occur to force us toward transition.

Who tells this story?

Academics like Timothy Mitchell and Roy Scranton;[2] disaster/apocalyptic narratives in popular culture (e.g. *Interstellar, Utopia*); numerous sf/dystopian writers, including Robinson, Margaret Atwood, and Paolo Bacigalupi;[3] those voicing a range of secular narratives of catastrophic transition, which are echoed in the eschatologies of religious communities.

What is the impasse?

The magnitudes of energy unlocked through fossil fuel use are what have allowed for the modernization of society. Every dimension of modernity is thus fundamentally dependent on the continuous presence of coal, oil, and gas. Technosocial lock-ins are reinforced by dominant political actors and hegemonic powers, and "naturalized" in civil society and everyday life. Then there are the enduring powers of infrastructure: pipelines, refineries, highways that push us to replicate behaviours and cultural forms. It is difficult, perhaps impossible, for the social majority to imagine and embrace a society that is not dependent on carbon energy. In turn, carbon political interests and agents lever their strategic discourses of power/knowledge. Transition thinking cannot escape the orbit of fossil fuels either (e.g. carbon capture and sequestration as salvation, discourses of energy security and energy equity). The dominant discourse and forms of infrastructure reinforce one another.

What is the pathway to action?

In this story, the current energy infrastructure maintains its dominance until the deterioration of the environment and lifeworld is so advanced as to produce some of kind of collapse or catastrophe out of whose ruins a transition might be born. The question of agency is a murky one. Either "we're fucked," as Scranton writes, or perhaps we simply won't be able to comprehend the path to transition until our energy infrastructure itself changes from below. In any case, in this narrative the artist or thinker plays a key role in speculating about the possible futures that could emerge out of collapse or in illuminating how we might live ethically with these catastrophic possibilities in mind.

Lingering questions?

Are we convinced that the artist/intellectual matters in this context? How can we tell these stories in a way that people find generative and engaging, rather than alienating and fearful? How do we frame the "unimaginable"? Is there value in considering the consequences of current energy impasse as "unimaginable"? Is there risk in advocating for dwelling rather than for action? Is there a useful "utopian" counter-narrative to the dystopic or catastrophic one?

Conclusion

These six different stories about what comes after oil, and all the many variations they take in our public and private lives show us that the questions around what to do about our carbon dependency and its impact on the climate are complicated and, in many cases, contradictory. There is no one clear problem, nor is there one clear solution—if there were, we might possibly already be living *after* oil. However, in considering how we relate to these different stories, we can also consider how we relate to others who are invested in these questions and to the variety of impasses that are connected to the question: what comes after oil?

Through the process of assessing these six stories, we have come to realize that working to more specifically identify the variety of impasses that can arise in thinking through the transition to an after oil scenario is a motivating task. Rather than seeing these many problems and possible modes of action as evidence of an intractable impasse, we now view them as a useful set of tools to use in entering into the various conversations and actions that are happening around oil transition.

The Arts, Humanities, and Energy (or, What Can Art tell us about Oil?)

An energy transition will require us to move away from using fossil fuels to employing renewable forms of energy. But there's more to transition than substituting one form of energy for another. We will not make an adequate or democratic transition to a world after oil without first changing how we *think*, *imagine*, *see*, and *hear*. Since oil shapes our ideas and values as much as it does our infrastructures and economies, an intentional energy transition will require us to think anew about wealth, beauty, community, success, and a host of other ideas that form our societies and our selves. What better set of disciplines than the humanities—art, history, philosophy, cultural studies, religious studies, and so on—to help us grasp the history of our present and to imagine different possibilities for the future?

The arts and humanities are uniquely equipped to help us engage in a full, successful energy transition. How will they do so? To afford a full sense of the crucial role that the arts and humanities play in helping us transition away from fossil fuels, we provide a brief account of the distinct roles played by *words*, *images*, and *performances*.

Words

. . . academic research, novels, histories, poetry . . .

The arts and humanities provide spaces for individual and
collective reflection on the consequences of oil culture for
life on earth in ways that are more holistic and empathet-
ic than the ideas generated by corporate interests or the
24-hour news cycle. Within this space, we can think about
how to re-organize our societies so that they respond to
our needs more effectively, without trampling over nonhu-
man life forms and ecological processes that are essential
and valuable in their own right. One way to do this is by
philosophical or cultural critiques that force us to con-
front the inadequacies of our oil cultures. Another way is
through speculative fiction that imagines what a post-oil
world might look like. As authors Margaret Atwood and
Neil Gaiman have recently argued,[1] now that we have the
facts about oil and climate change, we need fictions to act
on them.

The arts and humanities also create knowledge that can
help us to see that social change is desirable and possible. Scholars create knowledge, in part, by revealing and
critiquing the ideologies that shape our notions of what
constitutes knowledge, beauty, common sense, and the
common good. More recently, they have sought to specify
the ways in which particular energy regimes impact our
perceptions, bodies, and communities. Art can similarly,
in the words of philosopher Jacques Rancière, "redistribute the sensible" and help us to relate energy to our
social experience beyond the standard language of energy "problems" and "solutions" that has so far failed to
achieve meaningful change.[2] We need to perceive the
world differently in order to change it.

Historical consciousness is a crucial ingredient for

robust collective reflection that alters how we perceive the world. The humanities can re-narrate the histories of oil and energy to enable us to think more fully about our current circumstances and future possibilities. Such histories can reveal the hidden or obfuscated traumas of the past that continue to shape our societies or that should guide how we proceed. For example, our sense of the our overall historical "progress"—the steps that we have taken in order to become modern—looks different when we understand that crucial role played by greater and greater access to, and use of, energy; progress and energy use have not only gone hand-in-hand, but the latter has had a determinate impact on the former. Telling stories about the past is not just an exercise in uncovering lost causes, traumas, and oppression. It can also point to alternative ways of thinking and being that may have been forgotten or suppressed in the mad rush to cover the world with oil.

Language enables thought, which means that we need to do the work of creating languages and bodies of knowledge that will help us to understand the world anew. The work of many artists and humanities scholars shows that the concepts that we use to understand our world have histories that shape their meanings. Intriguing possibilities open up when we question the concepts that we take to be essential and seek to create new ones that enable new meanings. One particularly demanding concept of late is that of "the economy," which the media discusses as if it were a living entity that makes claims on us. We are told that must organize ourselves and behave in ways that are "good for the economy," and we want politicians to "manage the economy" effectively. But as the scholar Timothy Mitchell has recently shown, the idea of "the economy" as an entity unto itself requires cheap energy to exist, and only acquired its current meaning in the early twentieth century.[3] Mitchell suggests that we could have an economy that is not structured around growth, as we currently do, because "the economy" is not a transcendent reality with a fixed nature. Other scholars have made similar analyses of ideas like "Nature" and "energy," both of which also have histories and may also be fostering unsustainable ways of being in the world.

On the other hand, the creation of new concepts enables new kinds of meanings. For instance, scholars and artists have begun to talk about "petrocultures" and "the Anthropocene," which are powerful ideas with the capacity to reshape how we think and talk about the world. If we specify our culture according to energy resources via the concept of petrocultures, might we not open up other ways of imagining our social existence? What other sort of culture might we want to create—a wind culture, for instance, or a culture of renewables? And how might the notion of humanity as a geological force, expressed in the concept of the Anthropocene, change how we see our world and our activities in it? Art and the humanities help us to see that life becomes possible in different ways in relation to how we use concepts; they equip us to think carefully about the kinds of concepts that we want to use, and why.

Making an intentional and democratic energy transition is a difficult task, in part because we are implicated in the system that we are committed to changing. Powerful oil companies and others opposed to change know this difficulty well and exploit it frequently. But humanities scholars and artists can help us to persist by fully coming to terms with the social challenges that we face and the possibilities that could lie beyond them.

Images

... films, paintings, visual arts, television, advertising ...

Like words, images can help us to think differently about the world as it is and as it could be. But images can do other things, too. Our society is saturated by images, which can sometimes feel oppressive by their sheer number alone; but images can also be subversive and liberating. Images often circulate more easily than words and traffic more effectively in meaning. They inspire strong moral and emotional responses, which can turn them into powerful symbols for ideas, movements, and beliefs. Complex ideas and human experiences can be distilled powerfully into a single image. Of course, images are frequently reinterpreted and appropriated to serve corporate or state interests, but they can also undermine dominant interpretations of the world, while offering new meanings to replace them. Images can fundamentally alter our perception of our world.

The image below is a reproduction of a piece by artist (and AOS researcher) Ernst Logar. When we look at this piece as artists and critics, the word "requested" jumps out at us. What does it mean to *request* energy? *Requiring* energy connotes necessity and utility (as in: how much energy does one *need*)? Had Logar used the word "requiring"—the verb that might more commonly be stuck into this sentence—the process of getting energy disappears. But *requesting* energy poses a different question: what are the social relations that lie behind this energy and make it accessible to us? Who is doing the requesting? And why?

One of the central ambiguities of Logar's piece concerns the idea of energy itself: whose energy is this? Oil energy? The artist's energy? The energy industry equates fossil fuels with all the good things of modern life.

The energy requested to write these words is not infinite.

Ernst Logar. "The energy requested..." (crude oil on paper, 2014)

Life wouldn't be anything, it suggests, without "energy."
Artists and scholars are interrogating these assertions, ask-
ing: when did energy become an abstract idea? What other
kinds of life become possible after oil? Logar's work distils
these complex conversations and these questions into one

provocative image.

Energy corporations understand the power of images to distil ideas and create impressions with emotional power. Enbridge has launched a marketing campaign that capitalizes on the vagueness of our notions about energy to insert itself into the most intimate and memorable experiences of our lives. For Einstein, E= was a mathematical formula. For Enbridge, E=life itself. Our social and personal lives, the ads say, are only possible with the energy that Enbridge provides. Dinner with dad; making memories; doggy smiles; warm welcomes; and guilty pleasures. Enbridge uses these expertly crafted images to tell us that happy and fulfilling lives depend on them.

But art can be put to purposes other than corporate interests. Below is an image that we developed in a playful Adbusters moment. It shows how artists can help us to see both that Enbridge's easy equation of oil energy to life is untrue, *but also that such an equation is central to our way of life now.* Life has been limited to life within a petroculture. This visual also helps show how creative protest includes resistance to the appropriation of creative rhetoric. It suggests the power of images to disrupt savvy marketing by revealing other truths. In one of Enbridge's E= equations, we see an image of a car driving on a windy road, along a rugged seaside of forested mountains.[4] Gorgeous! What one might not grasp from the Enbridge ad is that the waterscape that we're photographing on our road trip is polluted with the very same fossil fuel that makes this vista so easily accessible to us!

We envisioned the image of transition to "after oil" as partly an issue of visibility that we approached in terms of an archetype. We asked ourselves, if oil is the personal unconscious of modernity, then how do we make unconscious energy visible? Narratives and visual narrative form

can make something visible and make the unconscious conscious so that we can grasp it and perceive it. Thus the fairytale of transformation can be seen as one archetypal narrative that captures the potential magic of oil and its transformative power, acknowledging oil's seductive qualities.

One example of a fairytale that we imagined is the story of Cinderella (a full version of this fairytale can be found on the AOS: www.afteroil.ca). In our version, oil is the magical power that crafts Cinderella's first transformation. We selected this classic in part because of the connection between the root of Cinderella's name—cinder—and fire or ash. Cinderella's life was forged by fire. This seems to us an apt connection to fossil fuels. Magic is the energy, the power, and the thing that can transform the mundane into something supernatural, just as it transforms Cinderella to who she is before the stroke of midnight. Oil is the magic that powers modernity. The power of oil is unconscious; we cannot grasp it and we don't perceive it.

In our version of the fairytale, Cinderella drives a white Audi whose energy is measured in horsepower. One horsepower was defined by James Watt as the amount of work a horse does to lift thirty-three thousand pounds of coal up the mineshaft one foot in one minute. At the stroke of midnight, the magic of oil wears off. The Audi transforms into a coach pulled by 220 workhorses. The image is, of course, absurd. Why does she need more than two horses? And where will she put all of her horses once she has used them to get her to where she needs to be?

In Cinderella the narrative of transformation—or the moral of the story—is about finding love and happiness. The magic certainly enabled her transformation and allowed her to achieve her dream. But when the magic is gone, what we realize is that happiness is not dependent on the magical powers created by oil. It is about the authentic connection of human beings to one another.

Performance

. . . action, drama, speech, performance art, events . . .

Everything symbolic and representational is always performative in some way. For example, the words and images to which we refer in this document are performative, because they produce meaning and identity through situated utterance. On one level, everything we say and do is comprehensible as performance, since we constitute ourselves in particular ways through speech, writing, and artistic creation. On another level, performance can consist of action, happening, or event—whether staged or not. Thus, when artists and scholars engage with the world through *performance*, we do so with the understanding that the term has multiple meanings across registers and disciplines. It refers to (1) the understanding of language as a process of producing meaning and identity; (2) the active and often embodied staging of dramatic or theatrical narrative; and (3) artistic work in which media and artists are organized around an event that is itself a signifying object or act.

Performance is commonly associated with liveness, improvisation, engagement, and process. It often reflects personal experience and the adverse effects of current systems, gesturing toward creative possibilities beyond them. In the context of thinking "after oil," performance can be understood as action that affirms the individual and collective right to imagine and inhabit a world that is not dependent on fossil fuels. Performance undertakes deliberate organized acts designed to present, problematize, and complicate our relationship as individuals and social groups to an oil-dependent world by:

» registering the effects of oil on human/animal bodies and communities;

» making visible the implications of these effects for the world we inhabit; and

» demonstrating how bodies in performance are both registers and agents of oil culture.

The linear temporal connotation of the phrase "after oil" underlines why performance is fundamental to the way we must think about this transition. "After oil" suggests that processes must be put in place in order to go from the "now" to the "after." And performance is itself processual: it simultaneously registers and responds, and so is always already in transition. Whereas words and images are *representational*, performance mobilizes *non-representationally* to encourage engagement with and analysis of the problematics to which words and images attest. In thinking about our position "after oil," we will find our place by taking words and images together and performatively working through them—potentially even demonstrating the transition itself in the process. Through improvisation, performance helps us practice how to get to "after oil."

Performance is often resistant-oppositional in nature, but it can likewise be reactive, interventionist, critical, revelatory, or productive. It can also combine any and all of these modes. Thus, we want to emphasize that forms of social activism in oil culture are always performative, but petrocultural performance does not necessarily have to be construed as social protest. Often, artistic performance invites *counterprocess* by calling on spectators to engage in active interpretation. As a result, it often registers an ambivalence that outright protest does not. The following examples, both of which merit equal attention and analysis, will illustrate the differences between petro-performance that is openly activist and that which opens up multiple interpretational possibilities.

Liberate Tate

In *Artwash*, Mel Evans describes "Liberate Tate," a petrocultural performance that renders its resistant-oppositional purpose deliberately unambiguous: the activist performance highlights the ways in which art gallery sponsorship obfuscates the damage oil can effect in its extraction and transportation.[5] Evans describes an event in which artists and climate activists entered Tate Britain to crash its annual summer party in 2010, which that year marked twenty years of British Petroleum (BP) sponsorship of Tate's UK art galleries. At precisely that moment, a blown-out wellhead owned by BP was expelling crude oil at the rate of 62,000 barrels per day into the Gulf of Mexico.

The artists and climate activists who entered the gallery staged two performances: in the first, they mingled with other guests before deliberately spilling ten litres of oil-like molasses, which they had been concealing under poufy skirts, on the polished stone floor of the gallery.

They then replicated the messy clean-up mission happening across the Atlantic: donning BP ponchos hidden in their handbags, they attempted to contain the spill while describing the mess to the crowd gathering around them as "tiny in comparison to the size of the whole gallery"—a dig at CEO Tony Hayward's initial (and widely criticized) defence of the BP disaster. At the same time, twelve more performers in black clothing spilled molasses from BP canisters at the entrance to Tate Britain, eliciting gasps from guests who continued to arrive at the party.

The artists and climate activists of "Liberate Tate" staged these two performances in order to protest and draw attention to the ongoing and catastrophic spill that BP was failing to resolve, and to point to the ways in which the enormously profitable corporation accumulated social and cultural capital (and so, too, moral standing) through the sponsorship of art. The performance of these artists and activists in the space of the gallery made visible the relationship between BP's self-congratulatory commercial operation and the mess they were making in public space—with far more catastrophic effects than the mess of molasses on the gallery floor. Performing radical protest while a BP party was happening made the important point that the company's sponsorship of Tate Britain and its art did not compensate for the effects of the spill and should not be counted as a sign of social and cultural responsibility.

Our Anaerobic Future

The intentions behind Aaron Veldstra's performance piece, "Our Anaerobic Future," are less explicit than the resistant-oppositional motivations of "Liberate Tate."[6] Using an archive of geographical data sets previously mapped for the purpose of oil exploration, Veldstra begins his performance by marking his wall-sized canvas (two sheets of drywall) with lines representative of pipelines, roads, and power lines in northern Alberta. He then retraces the map data using a syringe filled with dark Chinese ink reminiscent of crude oil. By the time he is finished tracing, his canvas is a sprawling palimpsest of blobs, beads, and drips. The initial lines only just discernible, the sections of drywall look like something Jackson Pollock might have created as a rebellious mud-logger during spare time on the rig. After tracing the last line, Veldstra sponges off the entire canvas using a combination of water and baking soda. Instead of throwing out the dirty water, he filters it through sand in a series of buckets. The next morning, he begins the entire process anew.

Depending on how one interprets this piece, it might be analyzed according to any of the modes described earlier. First, the insistence upon using ink in the place of crude oil demonstrates *resistance and opposition* to the unnecessary use of petroleum. Next, the refusal to waste water is *reactive*, *interventionist*, and *critical* of oil producers' attempts at remediation and sustainability. Lastly, the re-doubled lines on the canvas are *revelatory* and *productive* in that they demonstrate how the individual replicates the environmental damage created by oil extraction.

At the same time, though, Veldstra destabilizes all of these interpretations by literally erasing his piece every day, thus emptying it of the meanings and associations we take from it. Thus it refuses to remain attached to any single performative mode. When Veldstra's performance piece begins again, it is open to new interpretations and analyses. The piece, then, calls attention to its equivocality: as a performance, it is not the same as other forms of social and political activism, but it is not entirely separate from them either.

Word, Image, and Performance

At these multiple sites and through multiple forms, art and the humanities play an important role in the process of energy transition—and will continue to do. We need the insights of writers, artists and performers to help us imagine new ways of thinking, seeing, and living.

Energy Futures

Who gets to imagine energy futures?

Corporations, geologists, and engineers put a lot of thought and care into a future with fossil fuels. As artists, humanities scholars, and social scientists we offer something unique to help consider alternate energy futures. Moreover, conversations about energy transition create an opportunity to talk about broad social change in the world: economically, ecologically, politically, and socially.

Indeed, some of us are more beset by the compiled disasters of fossil capital than others. The road to the present has been a long one and its material legacies will continue to have profound, lasting effects. Even hundreds of years after oil, we will still be met with the hulking infrastructures of petromodernity. What's more, the carbon-dioxide saturated climate will continue to warm the planet with turbulent results for some time to come as the material, meteorological, and political disasters of fossil capital toss us back and forth like bits of plastic on the surf.

Confronted by the prospect of such legacies, our social systems buckle under the pressure of the need for change.

Energy futures can be more ethical futures. In addressing this task for the imagination, we insist on placing equal access to nutrition, water, shelter, healthcare, and education at the heart of how we imagine and enact energy transitions.

Today, we face the first energy transition in which we are globally and collectively aware. This energy transition and our energy future are socio-political projects, regardless of who oversees their development. Now is the time to make the collective decisions for a more just, more equal future, to insist on a guided energy transition that, at the same time, moves towards a future not only after oil, but after capital as well.

Part I: Infrastructures

Gridlife dependencies

When we imagine coal, oil, or the energy potential of wind, sun, or water, we presume resources that will work for us, toward some collective human good. After all, we (in the industrialized North at least) expect to flip a switch or turn the ignition key knowing that the power will be there. But this kind of gridlife is clearly not the same everywhere. Infrastructures are variable and changing, being developed or in ruination. Nearly 97% of those who live without electricity, about 22 million people, are in sub-Saharan Africa and Asia. A vast divide characterizes energy access; in the simplest terms there are those who expect to be *ever on* on the grid and those who have lived entire lives *being off* the grid. These are fundamentally different encounters with energy.

Our "addiction" to oil and electricity has become a truism. And addiction is a descriptive diagnosis because it suggests sickness and dependencies, (bad) habits and interventions. But unlike moral tales associated with the usual host of chemical dependencies—alcohol, nicotine, heroin—our energy crutches are not deeply questioned; at best our dependence is seen as simply a matter of swapping one form for another, or a plea to seek resources from one or another point on the planet. We rarely find ourselves questioning whether we s*hould need* energy resources; rather, we want to be assured that we can *have* them, whether in carbon or renewable forms. But what if energy were not always already "there for us"? What if we sobered up and broke that (now) deeply forged dependency?

Policy makers, engineers, and others are not likely to suggest that we go without or that we willingly stop—for parts of the day, or parts of our lives—indulging our energetic dependencies. But what if we in the global North were to be more like many of those living in the global South? What if we quit assuming a standing reserve of energy? In these times of transition and transformation, our aching reliance on energy, our electrical and chemical dependencies, must also be interrupted.

Centralized vs. decentralized energy infrastructures

If we prioritize equality when imagining our future with energy, how does this allow us to see energy infrastructure differently? Much of the discourse concerning the control of energy supplies is conceived in terms of a centralized vs. decentralized system—in other words, a state/corporate controlled power supply vs. a power supply generated by technologies owned by individual users. For instance, homes that access energy through power grids stand in contrast to homes off the grid that are self-sufficient and utilize an array of resource generating technologies.

However, in so far as decentralized energy systems are considered a response to larger structures of state and economic power, going "off grid" does not escape all of the conditions of petromodernity. For instance, the specialized technologies and materials that support off-grid homes often remain embedded in the larger material and economic economies of petromodernity for their construction and maintenance. Furthermore, modern grids assemble public and private sector labour that traverse different forms of governance. Finally, those who have the ability to go "off grid" but maintain petro-modern lifestyles (highly mobile, access to the full variety of available goods and services, access to a full range of information access) often represent a very privileged sub-set of the population.

It would be wise, then, to question the fantasies of off-grid living, which often involve privileged notions of indi-vidual autonomy, racialized visions of the wild as a place for whiteness, and an understanding of infrastructure as a self-contained set of materialities and practices. Instead, we need a more nuanced understanding of off-grid living, especially in the context of energy regime transition. Is the off-grid exodus in the industrialized global North (popu-larized by right-wing militia, left-wing urban bourgeoisie, and peak oil preppers), for example, an extension of white settler privileges, given the whiteness of existing off-grid settlements and trends (i.e. the tiny house movement) in the industrialized world? What are the differences between off-grid living in the industrialized world and the off-grid existences of those (many in the global South) who have never lived on a grid?

Part 2: Temporalities

Ways of seeing the future: prediction, vision, speculation, memory

Who can see the future and how do they claim to do so? Who has the right and/or the responsibility to imagine the future? Oil corporations such as Shell have asserted

that right. Pierre Wack from what was then Royal Dutch/
Shell claimed to have anticipated the dual oil crises of
the 1970s through a form of scenario planning, or what
is now known as *futurism*.[1] Shell's "Energy Scenarios to
2050" claims to predict the future with the same degree
of certainty.[2] Basing their predictions upon the exper-
tise of technocrats, scientists, and economists, they limit
energy futures to only two alternatives. But in seeing into
the future, these documents do not confine themselves to
"reasonable prediction." They put forward a "blueprint"
for the future, which also lays claim to visionary thinking.
This method of accessing the future might be imagined to
be the realm of the seer and the artist, but it is also rou-
tinely colonized by politicians and business leaders, who
have long since sought to tame "creativity" and to put it
to work in imagining and justifying a neoliberal worldview.

Far from offering a visionary account of a more just
energy future, documents such as "Energy Scenarios"
remain in thrall to the limits of what is imagined as possible in a world organized around the production and consumption of fossil fuels. They offer predictable manifestos
for a future after oil indebted to retaining and protecting
the values and desires of the fossil fuel age. Growth and
progress trump all other values; neither equality nor justice merit even a passing mention amongst the prescriptive
predictions and visions of the "Energy Scenarios." Since
the future is too important to be left to technocrats and
neoliberal leaders, other ways need to be found to gain
access to it. Tactics such as speculation or future-oriented memory offer opportunities for other voices to make
themselves heard. While "reasonable prediction" is based
on probability and a desire for certainty, speculation values
uncertainty; while "visionary thinking" reveals itself to be
rooted in the business-as-usual of the neoliberal present,

future-oriented memory invites the rediscovery of forgotten imagined energy futures. Both uncertainty and the rediscovery of forgotten energy futures offer scope for other voices to enter the fray and to place social and environmental justice on the agenda. Both break the frame of a single line connecting past, present and future.

The longue durée of petromodernity

"Democratic politics developed, thanks to oil, with a peculiar orientation towards the future: the future was a limitless horizon of growth," writes Timothy Mitchell in *Carbon Democracy: Political Power in the Age of Oil.* Instead of being an inevitable reflection of resource abundance, Mitchell argues that this perception of an energy future was "the result of a particular way of organising expert knowledge and its objects, in terms of a novel world called 'the economy'."[3] To envision the future of energy from a contemporary perspective—a perspective simultaneously from different geographies, economic and environmental conditions, and social matrices intersecting race, gender, and colonial relations—is, at the very least, to imagine an energy regime different from petromodernity but embedded in the durable legacies of petromodernity.

The petromodern society has produced legacies including global climate change and the near-ubiquity of durable waste such as plastics. Whatever is imagined as the ideal energy regime to follow that of oil, this orientation toward the future must necessarily be haunted by the long shadow of petromodernity's past. Some scholars have already provided useful terminology for engaging the future of a world in which the epoch of oil will have consequences for hundreds or thousands of years. For example, in *Slow Violence and the Environmentalism of the Poor*, Rob Nixon uses the term "slow violence" to describe violence that happens "gradually and often invisibly," an apt description of the environmental by-products of petromodernity such as oil spills, air pollution, nuclear contamination, and global warming. Timothy Morton uses the term "hyperobjects" to describe things that are massively distributed in time and space, and therefore difficult to describe or manage. Styrofoam, the Pacific garbage gyre, and uranium are all examples Morton gives of hyperobjects. These and other concepts are useful for imagining the *longue durée* of petromodern legacies.[4]

Multiplying temporalities

Is there an "after oil"? What time is it there? Whose time is it there?

Perhaps we are already in the *after*. We are in that *after*math of the dream/myth of economic progress, of defining a better or good life through accumulation with a great debt owed already to the future. In common economic language, we might say that the grandchildren's inheritance has already been mortgaged without their signatures. In this aftermath all beings, and all things, are always already and forever covered in oil. Thus, although contested, a word/concept such as Anthropocene might serve as interruptive or disruptive—a reminder that it is already and forever not business-as-usual; a reminder that this is not a time that is coming but one that is already here and now; and a reminder that this is our inheritance, and the inheritance of those to come.[5] This *is* a radical rupture in the capitalist line of progress and growth, revealing the latter as a misplaced and destructive narrative, both to humans and non-humans. The Anthropocene marks that "we" are already suffering the effects, side effects, and even future-effects (some of us more than others) of the oil (ka)boom.

What is this "after," when the future imagined by modernity has already passed and thresholds have been crossed? The incalculable loss through this carbon consuming present and future mass extinction event has inextricably altered what futures are possible. Any futurizing imaginaries and visions must take this loss into account. The petro-fuelled progress vision imagined always more; but there will always now be less. The extinction not only of species but also of myriad and diverse human cultures and languages, so intimately and intricately and sensitively entwined with what we call nature. This is an eco-bio-cultural extinction event, a homogenization, and a de-diversification. The names will be lost.

These questions summon up the thought of equality not just for the current inhabitants of the planet, but for the others to come—plants, animals and those ubiquitous grandchildren on whose behalf we dare not have faith that future technologies, that we do not yet know, will save them from the troubles created in this present (and *that* past). Equality means "consulting" the grandchildren's grandchildren today, and the honeybees and bumblebees and bats and moths, and future pollinators upon whose lives it all depends.

A frequently acknowledged paradox of the approaching collapse of industrial civilization goes something like this: on the one hand, the peak of global oil production represents a potential catastrophe for industrial civilization in which this ubiquitous resource would become less available for central activities such as transportation, agriculture, and manufacturing; on the other hand, if petromodernity persists beyond the current decade, it will likely ensure catastrophic global climate change and the extinction of most life on earth, including human beings. Gerry Canavan summarizes the potentially catastrophic paradox:

"That is: either we have Peak Oil, and the entire world suffers a tumultuous, uncontrolled transition to post-cheap-oil economics, or else there is still plenty of oil left for us to permanently destroy the global climate through continued excess carbon emissions."[6] This apparent paradox, in which either the continuation or discontinuation of petromodernity produces catastrophic circumstances for human communities, foregrounds the immediacy of the problem, the need to transition from fossil fuels to an alternative regime as soon as possible.

This is no apocalyptic vision

Energy futures tell us more about the present than they do about the future. Energy transition characterizes the global present, but the lived experience of that transition is not the same the world over and is characterized by inequalities on varying scales. To the protagonist of Mahmoud Rahmani's documentary *Naft Sefid* (*White Oil*), a lament for the passing of petroleum-fuelled optimism, the future is figured as loss. Those who have lost out are those who are left behind in the Iranian village when the extractive industries moved on, having exhausted the supply of oil in that location. They are left behind with the dust, the stones, and the wild dogs. This is no apocalyptic vision. This is the energy present for Rahmani's protagonists, for whom the future looks very different than it does, say, to a small contractor looking forward to the opening up of offshore oil reserves off Newfoundland, or to the corporate executive weighing up the dwindling reserves in the North Sea against the opportunities offered by the adventure of drilling in the Arctic, or to the urban slum dwellers in Lagos living off-grid not as a life-style choice, but out of necessity. Connecting these and many other diverse energy presents, however, is a prevailing sense of finitude. Oil is finite.

Part 3: Scale

Beyond scale as instrumentality

The concept of scale recurs repeatedly in discussions about energy futures. Energy transition is figured as massive and overwhelming, but also as unfolding in small, everyday ways. "Scale" is used in a multiplicity of ways in English. We talk about pay scales, the scales of a ladder, music scales, fish scales, scales on skin, scaling mountains, scales of justice, living life on a grand scale. We use the term to refer to a key to interpreting a map, as well as a device for measuring weights. Important in many of these uses of the term "scale" is a notion of comparative measurement, of assessing how various people and things fit into frameworks. In foregrounding equality then we must be pre-occupied with facilitating scenarios in which scales are balanced, in which resources and opportunities are distributed equitably. But how this should happen isn't self-evident. It might mean that the development of efficient and cheap off-grid infrastructures should be made paramount because they are more easily delivered and maintained by individuals and small communities. But thinking scale equitably might also entail constructing large-scale infrastructures to enable fairer energy distribution. It could also provide a justification for restricting the consumption of resources by those in the global North.

Other meanings of the term scale, ones without that sense of rational instrumentality, prompt us to think in quite different directions. Fish scales have nothing to do with notions of comparative measurement, but they evoke the existence of life forms that operate according to their own logics. There is a need to think about equality in a way that facilitates the coexistence of manifold forms of being—human, non-human, and post-human—and their various attachments to ecosystems in the world without placing these in hierarchies. We need to think beyond scale, in its instrumental sense, entirely.

What is wealth in a world after oil?

A holistic vision is necessary to enact energy transition that is equitable across cultures, geographies, and temporalities. Beyond gradual shifts in adapting alternate energy sources, the consistent rhetoric about the need for vast energy reserves for dependable delivery to consumers is one impediment to enacting alternative energy sources such as wind and fusion.

The possibility of off-grid options assumes the retention of grid networks and the impossibility of un-linked autonomous situations. The interdependency of ecological systems and acknowledged effects in the age of the Anthropocene undermine the isolated utopian situations that are arguably inflected with the gender, racial, and class attributes of privileged "settlers." The Anthropocene alerts us to the inequalities that persist between the global North and the global South. Energy transitions risk exacerbating those historic disparities.

Attention to the scale of the extraction and production of fossil fuels and to the possibility of their decreased availability for consumption in developing nations, coupled with an equitable redistribution of resources across nation states, is one strategy in addressing the destructive effects of fossil fuel. The unprecedented scale representing the fossil fuel economy, its culture and materiality, is incomprehensible and abstract in ways that create an impasse in addressing alternative cultural and material ways of living. Reciprocity and ethical actions that respect the non-human natural world for its limited capacity to provide for humanity are principles enacted by Indigenous peoples from whom we can draw relevant insights.

Modernity's promise and the capitalist imperatives that underwrite it is an increasingly unattainable measure of success. This does not mean its opposite, a return to feudalism or barbarism, is the other possible future. Enacting imaginative futures premised on embodied experience redefines the valuable, the possible, and the ethical. What is wealth in a world after oil? What should it be?

Conclusion

What we've offered here are new coordinates from which to imagine a successful, intentional energy transition, one in which technological and economic change is the *result of* collective social change (rather than the other way around). What, in the end, might we take away from the analyses offered here about the current shape of our petro-societies and the steps we should take to transition to societies no longer shaped and defined by fossil fuels? What issues and problems do we have to address and overcome to enable this transition—everything from shifting social habits and life expectations to undoing our dependence on many of the secondary products of petroleum (e.g., ink, tires, vitamin capsules, eyeglasses, footballs, detergents, parachutes, fertilizers, panty hose, aspirin, dyes, yarns, nail polish, plastics, dentures, bandages, linoleum, hair coloring, surf boards...)?

The thinkers who came together for the inaugural AOS were asked to answer four questions:

1. Considering historical precedence, what cultural strategies are available to trigger and expedite a large-scale transition of energy regimes?

2. How does the problem of energy force us to

rethink our traditional notions and categories of political agency?

3. How is the use of energy entwined with representations and narratives about modernity and the environment? Correspondingly, how do artistic productions reflect, critique, and inform our understanding and use of energy? and,

4. What range of scenarios is currently on the table for imagining our future with energy?

The key issue animating each of the above questions can be summarized in a single word:

1. Strategy

2. Agency

3. Representation

4. Futures

At a minimum, the analyses presented here are intended to make evident the multiple ways in which the forms of energy on which a society depends shape it in fundamental ways. This document reiterates the point about energy's fundamental qualities in each chapter in order to emphasize two related points. First, the optimism usually attached to renewables is that *they make the world made by oil possible after oil*, a failure of imagination we've sought to address. Second, while thinking the full picture of energy transition is tricky—keeping in mind the social, technological, economic, and environmental elements *in* transition—it nevertheless offers opportunities for large-scale change.

We have for too long been comfortable imagining energy—fossil fuels, in our own case—as a necessary, if generally unremarkable feature of human societies. We might well know that we need fuel to make our cars go, gas to heat our homes, and coal to generate the electricity that powers our high tech world.[1] However, the idea that oil, gas, and coal have had a determinate impact on the shape and character of our societies is not something about which we have been previously been aware. The analyses offered in *After Oil* point to the necessity of understanding how, where, why, and to what degree energy shapes and creates social belonging and individual being. We need to understand our societies as *oil* societies and our modernity as a *petro*-modernity to better grasp who and what we are. We also need to do so because we are entering a period in which we will undergo a transition from being oil societies to no longer being oil societies. Understanding how energy shapes society is essential to undertaking this transition, and draws attention to issues that we have avoided seriously addressing as we begin to engage in this unprecedented transformation away from a fossil fuel society.

While all four chapters remind us of the importance of energy to society, they also provide us with specific insights as to the direction and shape of our coming energy—and social—transformations. The "Principles of Intentional Transition" outlined at the end of the first chapter provide a series of principles about what we need to consider in order to transition out of our specifically economic dependence on energy (the process known as "energy deepening" detailed in that section). These *strategies* concerning a change in our relationship to energy include: equality of access to energy by people around the world, collective decision-making, ethically driven best practices about sustainable energy use, and a reimagining of how we comprehend growth and development. To evoke the title of Tim Jackson's book: we need to envision prosperity without growth.[2]

There is one further principle outlined in this chapter—
the one on which all the others are dependent. This is the
importance of *agency* in shaping an intentional transition.
This period of energy transition constitutes an opening
for substantial socio-political change unlike any encoun-
tered in recent memory. The need for an energy transition
isn't the result of a technical failure in our existing energy
systems, nor the outcome of the need for a response to
pressing environmental crisis of global warming. Rather,
the necessity for a shift in so fundamental an element of
modernity as the mechanisms that power it—materially,
socially, and even psychologically—constitute a judgement
on the principles around which we have shaped social life.
The transitions that will take place in coming years point
to the fact that we can't live the way we have lived, can't
organize ourselves in the way that we have organized ourselves, and can't fill our social imaginaries with the hopes,
expectations and beliefs that we have in the past.

Agency names that capacity for peoples and communities to collectively and consciously compose the way that
they want to live in this world. The recognition of the
role that energy has played in shaping social life to date,
and the need for a change to energy systems, means that
there is an opportunity for a significant alteration in how
we live, too. The incredible energy resources that many
(though certainly not all) people have enjoined over the
course of modernity have expanded their capacities and
opportunities to more fully enjoy and participate in a rich
and vibrant life. It has just as certainly created a situation
in which much of our life activity remains driven by a marketplace that measures its success by the index of profit
rather than quality of life and the health of individuals
and communities. An enormous opportunity will be wast-
ed if energy transition isn't accompanied by an equally

impressive social transition—one that allows our energy resources to enrich our lives, rather than exhaustively amplify our activities only to generate profit.

The narratives that drive our sense of transition—and so, too, our sense of agency in relation to energy transition—are the subject of the second chapter, "Energy Impasse and Political Actors." It is to be expected that there would be numerous narratives about the desired path that energy transition might take. Those who have benefited from the current energy system want the energy transition to take place in a manner that rocks the boat of contemporary power as little as possible; others see energy transition in the way we have suggested above—as providing an opening for political transformation that would redistribute the power embedded in political structures as much as in energy systems. In *After Oil*, we have identified six key narratives of energy transition, stories told by different social actors, with distinct ideas about the way to bring about change and the impediments to doing so. As we make clear, the point of identifying these narratives isn't finally to make a choice between them. Rather, this analysis of the ways in which the challenge of energy transition has been named and explained is intended to provide a deeper insight into the complexes of the current social landscape, including the sharp differences that exist around agency and the right way to move into a new energy future.

Narratives of energy transition are guided by distinct ideas of agency and pathways to change. They are equally shaped by the visions of energy *futures*. Transition requires a framing of a future toward which we are moving—a goal to be reached, a shift in how we live towards which we are reaching. And as Chapter 4 makes clear, the ways in which these futures are figured—prediction, vision, speculation,

and memory—matters as much as the after-oil scenarios
that are painted. The fantasy of limitless growth that has
long given life to capitalism is today hemmed in by eschatologies that mark endings and beginnings; these borders
of time speak to the present and our sense of power and
social possibility, as much as they do to the futures they
name. Any contemplation of energy in relation to the
future highlights one of the biggest changes we will have
to make alongside a shift in the energy we use. Energy
has been connected to wealth throughout modernity, both
through the sheer value that it has added to economies and
the process of energy deepening through which expanded
energy use and expanded wealth have become synony-
mous. One of the challenges posed by energy futures is
the need to rethink those basic measures of value that we
have been told repeatedly to leave alone: GDP, profit, and
growth. These are social inventions like any other; our
present energy transition might be the time to cast them
aside as categories that are no longer doing anything other
than getting in the way of human progress.

Fossil fuels are at one and at the same the most material
of substances, dragged dripping from the soil and shunted
along pipelines from one spot on the earth to another,
and also the stuff of fantasy, sheer potential that can be
actualized for creative as well as destructive purposes.
Energy transition reminds us that the societies we have
shaped around fossil fuels are collective fictions. There is
no necessity for society to have taken the shape that it
has, just as there is no necessity for it to continue to have
this same form: social life isn't fate but a world shaped by
those within it. The struggle that is currently taking place
over the direction of energy transition, which involves
scientists, activists, governments, and businesspeople, is a
struggle over representation and narrative, the stories we

tell about human capacity and future possibility. Those of us involved in the *After Oil* project will often turn to a mantra when it comes to the basic rationale of our project: while scientists may have definitely told us about the reality of global warming, they've given us no clue as to the path forward from the present to the energy futures we want. This is why the input and energies of the arts, humanities, and social sciences are crucial to energy transition: they give us insights into the *representations* that have guided our imaginings and those that might yet lead us into a future after oil, one even more full of possibility than the one we are leaving behind.

After oil: the phrase can sound like a threat or the naming of an apocalypse. This project will have accomplished its intent if "after oil" changes its valence, becoming the name for a place and time in which we want to be.

Notes

Notes for Chapter 1: Triggering Transition

1. "Cruel optimism" is a phrase used by Lauren Berlant to capture the affective, emotional dynamics that have stifled contemporary political change. For Berlant, contemporary narratives of a better future through social change generate an optimistic belief in the "possibility that the habits of a history might *not* be reproduced." However, they do so in a way that inhibits us from actually undertaking this change. Despite significant problems with the way we live today, publics tend to "choose to ride the wave of the system of attachment that they are used to" instead of leaping into a new way of living. See Lauren Berlant, "Cruel Optimism," *differences: A Journal of Feminist Cultural Studies* 17.3 (2006): 31, 23.

2. "Energy deepening" names the process through which economic growth becomes dependent on ever-increasing quantities of (non-human) energy. For a discussion of energy deepening, see Bernard C. Beudreau, *Energy and the Rise and Fall of Political Economy*, Westport, CT: Praeger, 1999. For an analysis of the aesthetic and cultural implications of energy deepening, see Jeff Diamanti, *Aesthetic Economies of Growth: Energy, Value, and the Work of Culture After Oil*, PhD dissertation, English and Film Studies, University of Alberta, 2015. See also the section "energy deepening" later in this document.

3. Andreas Malm, "The Origins of Fossil Capital: From Water to Steam in the British Cotton Industry," *Historical Materialism*

21.1 (2013): 31.

4. See Vaclav Smil, *Energy Transitions: History, Requirements, Prospects*, Santa Barbara, CA: Praeger, 2010.

5. See, for instance, Yannis Stavrakakis, "Objects of Consumption, Causes of Desire: Consumerism and Advertising in Societies of Commanded Enjoyment," *Gramma* 14 (2006): 83–105.

Notes for Chapter 2: Energy Impasse and Political Actors

1. Robinson's Science in the Capital series includes *Forty Signs of Rain* (2004), *Fifty Degrees Below Zero* (2005), and *Sixty Days and Counting* (2007).

2. Timothy Mitchell, *Carbon Democracy: Political Power in the Age of Oil*, London: Verso, 2013; Roy Scranton, "Learning How to Die in the Anthropocene." *New York Times*, November 10, 2013.

3. In addition to Robinson's Science in the City trilogy, see, for example, Margaret Atwood, *MaddAdam*, Toronto: McClelland & Stewart, 2013; and Paolo Baciagalupi, *The Windup Girl*, San Francisco: Night Shade Books, 2009.

Notes for Chapter 3: The Arts, Humanities and Energy

1. See Atwood, "It's Not Climate Change—It's Everything Change." Available at: https://medium.com/matter/it-s-not-climate-change-it-s-everything-change-8fd9aa671804#.bqsq0sryo; and Gaiman, "In conversation with EIA: Neil Gaiman on the natural world." Available at: https://eia-international.org/in-conversation-with-eia-neil-gaiman-on-the-natural-world.

2. See Jacques Rancière, *The Politics of Aesthetics: The Distribution of the Sensible*, trans. Gabriel Rockhill, London: Continuum, 2004.

3. See Mitchell, *Carbon Democracy*.

4. Examples of the Enbridge's "Life takes Energy" ad series, which features the "E=" motif, can be found at: http://www.enbridge.com/AboutEnbridge/Life-Takes-Energy.aspx.

5. Mel Evans, *Artwash: Big Oil and the Arts*, Chicago: Pluto Press, 2015.

6. A video of the Veldstra project can be found at https://www.youtube.com/watch?v=XNPOOMYfZM0.

Notes for Chapter 4: Energy Futures

1. Pierre Wack, "Scenarios: uncharted waters ahead," *Harvard Business Review,* Sept-Oct 1985.

2. Available at: https://s00.static-shell.com/content/dam/shell/static/future-energy/downloads/shell-scenarios/shell-energy-scenarios2050.pdf.

3. Mitchell, *Carbon Democracy*, 253; 142–143.

4. Rob Nixon, *Slow Violence and the Environmentalism of the Poor*, Cambridge: Harvard UP, 2013; Timothy Morton, *Hyperobjects: Philosophy and Ecology after the End of the World*, Minneapolis: U Minnesota P, 2013.

5. Atmospheric scientist Paul Crutzen and ecologist Eugene Stoermer coined the term "Anthropocene" to capture the impact of human activities on the planet. It is the proposed name for the present geological epoch (following the Holocene) and highlights the degree to which humans have reshaped the Earth's environment. While there is disagreement about the precise beginning date of the Anthropocene, its length is understood to be a few hundred years—not the thousands of years that geological epochs typical demarcate (e.g., the Holocene is 11,700 years long).

6. Gerry Canavan, "Introduction: If This Goes On," *Green Planets: Ecology and Science Fiction*, ed. Gerry Canavan and Kim Stanley Robinson, Middletown CT: Wesleyan UP, 2014, 5.

Notes for Chapter 5: Conclusion

1. Coal-fired power capacity has increased 75% since 2000; it now supplies 41% of electricity on the planet. See Eric Reguly, "Key questions as negotiators mark third day of Paris climate summit," *Globe and Mail*, December 3, 2015: A9.

2. Tim Jackson, *Prosperity without Growth: Economics for a Finite Planet,* New York: Routledge, 2009.

Cover design by Jonathan Dyck.
Interior design by Jen Hedler Phillis.
Text set in Garamond by Robert Slimbach.
Headlines set in Futura by Paul Renner.